Contents

Contents

COMPLETED

5 Is the geography of Russia a curse or benefit?

1.1 What is a geographer?

Geography happens every minute of the day

1 Find a news report in a newspaper or from a website and stick it in the space below.

2 Describe what the story is about. You may wish to use the questions to help you.

Where is the story set?
What is this place like?
What happened at that place?
How is that place changing/has it changed?
Who is/was affected by the changes?
How does the story make you feel?

I think it is geographical because _____

_____

3 Is your story mostly about the **physical world** – the work of rivers, seas, ice, or weather and climate; the **human world** – where people live and work; the **environmental world** – habitats?

I think the story is mostly about the _____ world because it involves _____ .

I think the story is about both the _____ and the _____ world because it involves _____ .

I think the story is about how **physical / human** world activities affect the environment because it involves _____ .

1.2 How has our knowledge of the world progressed over time?

Mapmaking in the past, in the present, and in the future

1 Study the two images below. Map A was created by Eratosthenes 2300 years ago and the satellite image A was taken in 2000. Draw arrows to the following modern day areas on **each** image.

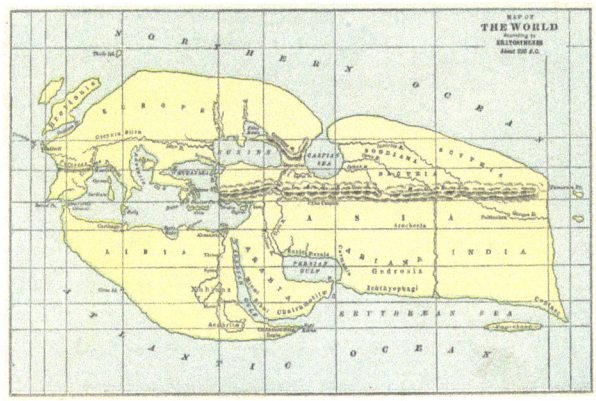

Map A: Eratosthenes' map

a The British Isles

b Scandinavia

c North Africa

d Italian Peninsula

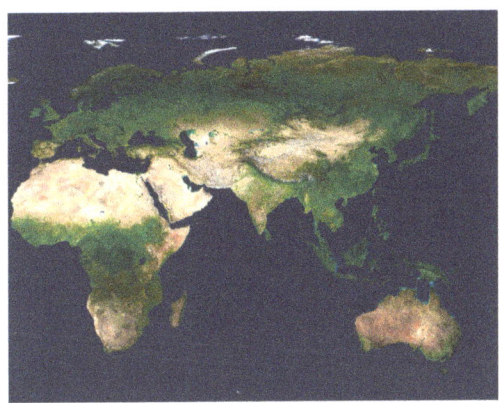

Image A: Satellite image

2 a) What do you think is the most useful aspect of Eratosthenes' map?

_____

_____

b) What is the biggest disadvantage you would have using it?

_____

_____

3 Go to www.google.co.uk/maps. How could you use each of the following features when investigating an area?

a) Map view	b) Street view
c) Zoom	d) Satellite image

1.3 What locational knowledge do you have of the world?

Locating the major physical features of Europe

1 Complete the table by naming the country where you can find these mountain ranges. If the mountain range is in more than one country, name at least two of them.

Mountain range	Country/countries
The Scandinavian Mountains	
The Ural Mountains	
The North West Highlands and Grampians	
The Pyrenees	
The Alps	
The Apennines	
The Carpathians	
The Caucasus Mountains	
The Balkan Mountains	Bulgaria and Serbia

European countries bordering Russia

1 Make a list of the European countries that have a land border with Russia.

Map A: Europe

2 Complete these sentences.

a) Russia's border with Poland and Lithuania is unusual because _____

_____

b) Other countries, like Turkey and Sweden, do not have a land border with Russia but are a

short distance away across the _____

and _____ seas.

© Hodder Education 2018

1.4 How can we locate places around the word?

Lines of latitude and longitude on world maps

1 Circle the correct words to complete the following sentences.

a) The imaginary lines that circle the Earth from side to side are called lines of **latitude** / **longitude.**

b) The imaginary lines that circle the Earth from top to bottom are called lines of **latitude** / **longitude.**

c) Lines of latitude show us how far **north or south** / **east or west** a place is from the Equator.

d) Lines of longitude show us how far **north or south** / **east or west** a place is from the Prime Meridian.

> Each degree of latitude or longitude can be divided into 60 minutes. 30 minutes is the same as half a degree. Antananarivo's full latitude and longitude in an atlas might read 18 52S 47 30E.

2 Look at Map A below.

a) Label the main lines of latitude: the Equator [0°], the Tropic of Cancer [23½° North], the Tropic of Capricorn [23½° South], the Arctic Circle [66½° North] and the Antarctic Circle [66½° South].

b) Label the main line of longitude: the Prime Meridian [0° East or West].

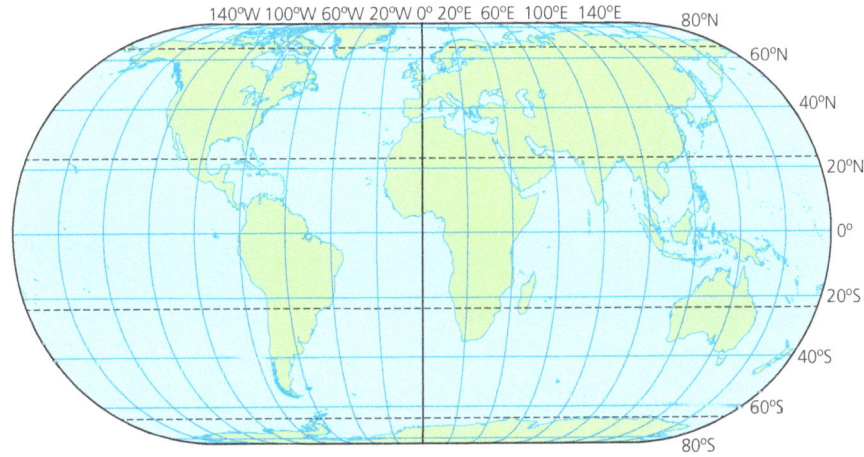

Map A: Lines of latitude

3 Look at Map B.

a) Carefully label: the Equator, 20° North and South, 0° – the Prime Meridian, 20° West, and 20°, 40° and 60° East.

b) Which of the cities marked A–D is at 0° North 32½° East? On the map label it **Kampala**.

c) Which of the cities marked A–D is at 9½° North and 14° West? On the map label it **Conakry.**

d) Which of the cities marked A–D is at 0° North and 9° East? On the map label it **Libreville**.

e) Which of the cities marked A–D is at 19° South and 47½° East? On the map label it **Antananarivo.**

f) On the map label **Accra**, which is at 5½° North and 0° West.

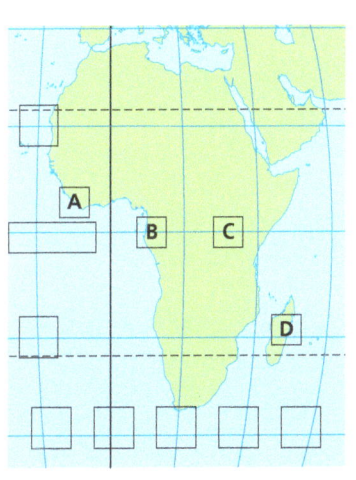

Map B: Africa

1.5 Why do we use OS maps to investigate places?

How maps represent the real world using scale

1 Draw lines to match the names to the way scale is represented on a map.

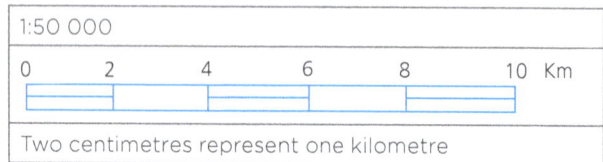

A statement of scale	1:50 000
Ratio scale	0 2 4 6 8 10 Km
Linear scale	Two centimetres represent one kilometre

2 Complete the table below to show how large scale and small scale maps are different.

	Large scale map	Small scale map
Area		A large area is covered on one map
Buildings/ features	Individual building shapes and field outlines shown	
Ratio scale		Usually 1: 50 000 or 1: 250 000
Uses	Planning	

3 Study the road map below.

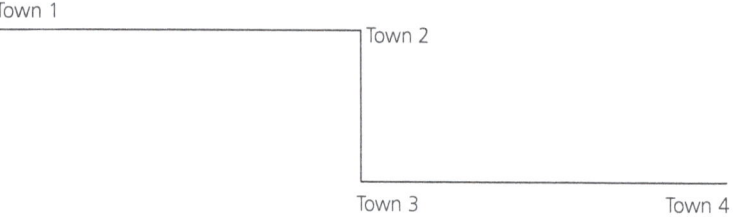

Town 1
Town 2
Town 3   Town 4

a) If the map linear scale is 1 cm represents 1 km, how far is it from Town 1 to Town 4 by road? _____

b) How far is it as the crow flies from Town 1 to Town 4? _____

4 Study the road map below.

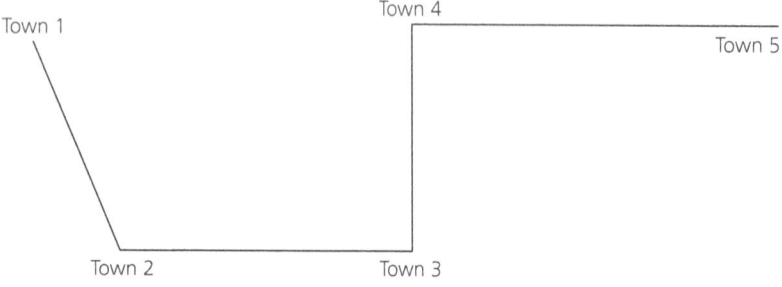

Town 4
Town 1
Town 5
Town 2   Town 3

a) How far is it as the crow flies from Town 1 to Town 4? The map scale on this map is 1: 50 000 or 2 cm represents 1 km._____

b) How far is it from Town 1 to Town 5 by road?_____

c) How far is it as the crow flies from Town 1 to Town 5?_____

d) If a new road linked Town 2 to Town 4, how much shorter would the journey from Town 1 to Town 5 be?_____

Can you locate features using map symbols and grid references?

1   Download and print a copy of the symbols for 1:50 000 and 1:25 000 OS maps from the Mapzone website (click 'Map skills' then 'Resources and links'). Use your copy of the 1: 50 000 symbols to help complete the following exercises.

2   Look at this map. Draw symbols to complete the key, using your symbol chart to help you.

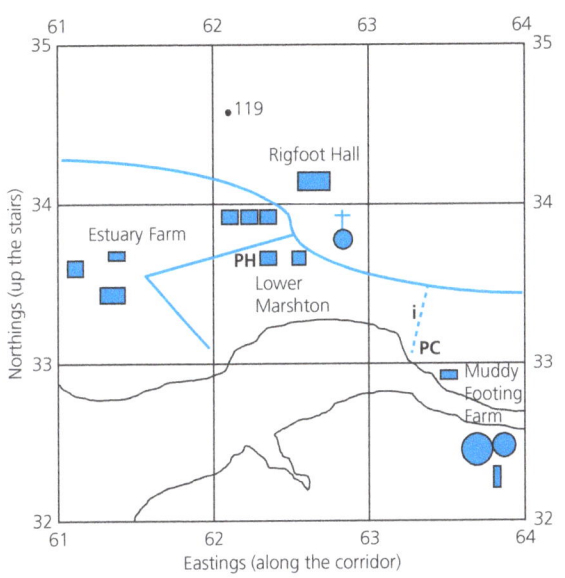

Key

Church with a spire
Information point
Toilets
Public house
Spot height

3   Give the four-figure grid references showing where you would find:

   a) The village of Lower Marshton _____

   b) Muddy Footing Farm _____

4   What feature is found at each of the following grid references?

   a) 634 331 _____

   b) 628 337 _____

5   Give the six-figure grid reference for the following:

   a) the public house _____

   b) the 119-metre spot height _____

   c) the information point near Muddy Footing Farm _____

6   Now complete the map by adding symbols for the following features, and add them to the key.

   a) Brow Top Woods, a woodland in grid square 63 34

   b) Sea View Caravan Park, at grid reference 617 322

   c) Boggy Edge Marshes, at grid reference 623 324

   d) A telephone, at grid reference 630 335

1.7 How do OS maps show height, direction and slope?

Using contour lines, spot heights and direction to describe landscape features on a map

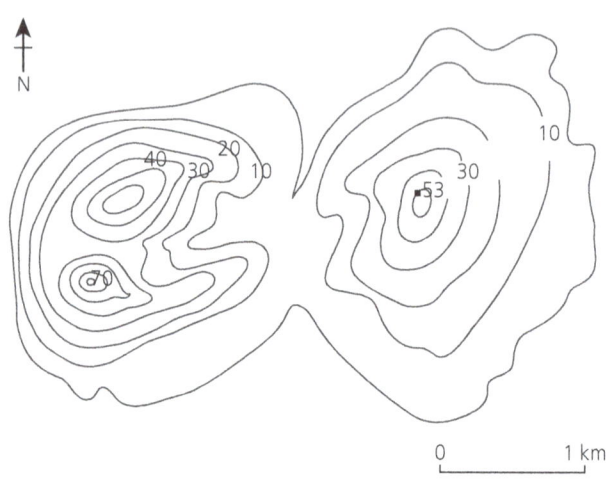

Map A is a contour line map of Three Hill Island. Contour lines join places of the same height above sea level. The contour lines are at 10 metre intervals.

Map A: Three Hill Island

1 Add these labels to the map.

| 20 | sea level | 40 | steep slope | 50 | gentle slope | 60 | spot height |

2 **a)** What can you say about the height of the north-western hill?

The north-western hill is above _____

b) Why can you not give the exact height?

I cannot give the exact height because contour lines _____

_____

c) What is the highest point on the eastern hill?

The highest point on the eastern hill is _____

d) How do you know this is the highest point?

This is the highest part of the eastern hill because _____

_____

3 Write down the height of the land:

a) 0.5 km **north-west** of the spot height on the eastern hill _____

b) 1 km **north-east** of the spot height on the eastern hill _____

c) 1.5 km **south-east** of the spot height on the eastern hill _____

4 Think of a more imaginative, but geographically suitable, name for each of the hills based on its shape and gradient. Write these on the map.

1.8 How can we use aerial photos with OS maps?

Using aerial photos to create maps

| directly | above | below |

1 Complete the following sentence by filling in the blanks.

A vertical aerial photograph is taken from _____ _____ the subject.

2 Find a copy of a vertical aerial photograph of your local area, using an internet search or Google Maps. Stick your photograph in the space below.

a) Add labels to your photograph to show the land use in your area. You may wish to use some of these words:

Farm	Farmland	Harbour	Housing	Industry/Industrial area
Marshland	Moorland	Motorway	Park	Port
Road	River	Rural area	Urban area	Woodland

b) Use the information on the photograph to create a land use sketch map. Make sure it has a title, a scale, a direction arrow, a frame and a key.

Key

3 Write down one advantage and one disadvantage of using a vertical aerial photograph to draw a map like this.

Advantage _____

Disadvantage _____

1.9 How do you investigate a locality by conducting fieldwork?

Preparing for successful fieldwork

1 Think about your journey to school each day. Shade in any of the features you pass on your walk or car/bus/train journey. Add any feature not mentioned in the spaces provided.

Offices   Parks   Terraced houses   Flats   Tower blocks

Houses with big gardens   Factories or workshops   Shopping streets or centres

Leisure facilities   Single shops   Streets with lots of parked cars   Busy roads

Open fields   Residential areas   Woodland   Villages   Rivers or streams

Coastline

2 Choose one residential area and two other areas you could investigate for a class fieldwork walk. For each area generate at least two questions that would help you describe that area and write these in column A. Generate another two questions you could investigate on the fieldwork and write these in column B.

	A What is it like?	B Where is it, why is it there, how is it changing, who uses it, how does it make me feel?	
The park	What play equipment is there? Are there lots of benches? Is there a litter problem?	How close is it to the residential area and schools?	Does this feel like a good area for me to use for recreation?
1 Residential area			
2			
3			

3 Investigating geography outside has safety hazards. Think of three considerations you need to make on your route to keep your class safe.

Safety point 1 _____

Safety point 2 _____

Safety point 3 _____

1   Study Image A. On the photo, label and name:

   a) two aspects of the physical geographical world

   b) two aspects of the living environmental world

   c) three types of human land use.

Image A: Ground level photograph of the Weaver and Mersey Rivers

2   Complete the sentences to answer the questions.

   a) Describe one advantage ground photographs have over vertical aerial photographs.

     One advantage of ground photographs is _____

     _____

   b) Describe one advantage of vertical aerial photographs for geographers.

     One advantage of vertical aerial photographs for geographers is _____

     _____

3   a) If contour lines on a map are drawn close together, does this represent a steep or gentle slope?

     Contour lines drawn close together represent a _____

   b) If contour lines on a map are drawn wide apart, does this represent a steep or gentle slope?

     Contour lines drawn wide apart represent _____

   c) If the height of the area around the wind turbines was drawn on an OS map would the contour lines be close together or far apart?

     They would be _____

4   If you were locating the following places on an OS map, would you use a four- or a six-figure grid reference? Circle the correct answers.

   a) A village   **four-figure / six-figure**   b) A post office   **four-figure / six-figure**

   c) A railway station   **four-figure / six-figure**   d) An area of   **four-figure / six-figure**
                                                                       woodland

5   a) Name two things a map should have which the photograph lacks.

     1 _____   2 _____

   b) Choose one of these things and say how it helps to answer geographical questions.

_____

2.1 How do we use our planet as a natural resource?

The Earth's different spheres produce natural resources

1 Read the following statements and write whether they are true or false.

	TRUE	FALSE
Earth is one of several planets in the solar system that can support life.		
Earth has the correct natural resources for life to exist.		
Some things are essential for our survival; these are called raw materials.		
Humans use raw materials to make things to meet our needs.		
Planet Earth has four spheres; the atmosphere, the hydrosphere, the lithosphere and the biosphere.		
The biosphere is all living matter.		
Water is part of the lithosphere.		
The atmosphere is a fragile layer of gases.		
Rocks are part of the lithosphere.		

2 If you have written that a statement is false, write the correct statement in the space below the statement.

3 Complete the following two sentences by circling the correct answers.

   a) **Renewable resources** are resources that can be replaced by a natural/manmade process in more/less than a human life span.

   b) **Non-renewable resources** are resources that cannot be replaced by a natural/manmade process in more/less than a human life span.

4 Study the list of resources below and place them in the correct column in the table. Add at least one more example of your own to each column.

| Coal | Gas | Wind | Oil | Air | Water | Gold |
| | Diamond | Timber | Chalk | Sand | Sunlight | |

Renewable resources	Non-renewable resources
_____ _____	_____ _____
_____ _____	_____ _____

2.2 What are rocks and how are they a natural resource? Part 1

2.2

Student's Book pages 24-25

Natural resources are created over millions of years

The Earth's history is often compressed into an imaginary day. Each hour represents roughly 192 million years.

1 On the timeline add the number of million years ago for

   a) 15:00 hours

   b) 06:00 hours

   c) 00:00 hours.

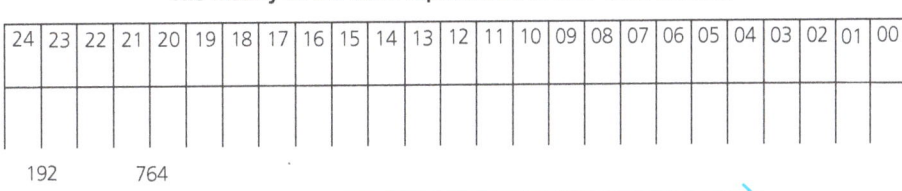

The history of the Earth represented as a 24-hour timeline

24	23	22	21	20	19	18	17	16	15	14	13	12	11	10	09	08	07	06	05	04	03	02	01	00

192   764

Million years ago

2 Draw lines to the timeline using the 24-hour scale to indicate when

   a) the first humans developed (23:59)

   b) mammals developed (23:39)

   c) dinosaurs developed (22:56)

   d) coal was formed in the Carboniferous era (22:25)

   e) the oldest fossils formed (05:36)

   f) the Earth's formation occurred (00:00)

3 Which type of rock formed first? IGNEOUS / METAMORPHIC / SEDIMENTARY

4 Weathering by exposure to the Earth's other spheres breaks down rocks in the lithosphere. Complete the flow charts to show how the processes of freeze–thaw and chemical weathering operate. Use the phrases below to complete the boxes.

FREEZE–THAW weathering

CHEMICAL weathering

Water runs into cracks in a rock. / Carbon dioxide in the air makes rainwater slightly acidic. / Repeated freeze/thaw cycles weaken the rock. / The rocks are dissolved especially at cracks and joints. / Eventually the rock is broken down to scree. / The weak acids react with limestone and chalk. / Water freezes and expands by 9%, exerting stress on the rock. / Acidic rain water falls onto rock surfaces.

2.3 What are rocks and how are they a natural resource? Part 2

The Earth's natural resources used as raw materials

1 Find a photograph or make a field sketch of a building in your local area. It could be your house, your school or an interesting older building. Glue or draw it in the space below and give it a title.

2 Annotate it by adding labels to show how it uses the Earth's natural resources as raw materials.

3 Look at Image A. Identify two raw materials used in the street area and buildings in Gold Street, Shaftesbury that are the same as in your photograph and two raw materials that are different.

Similarity 1: _____

_____

Similarity 2: _____

_____

Difference 1: _____

_____

Difference 2: _____

_____

Image A: Gold Street, Shaftesbury

4 Write down any materials in the buildings which come from renewable resources.

_____

_____

2.4 Why are soils the root of life?

1   Complete the paragraph about soils with the words from the box below. You need to unscramble the words before you use them.

spouor	streinunt	perlbeimeam	lsrenima	kbdecro
risdeb	cinagor	ratnual	shumu	tbcariea

Healthy soils are vital to give mineral _____ and water to plants' roots. Plants form the basis of the food chains we need to survive so soil is an important _____ resource. Soil is composed of layers of _____, water and organic matter from both a _____ layer underneath and a top layer made of decaying plant and animal _____. Organisms like fungi, worms and _____ break down the organic material to form _____ and mix it with weathered rock minerals to form the surface and subsoil. Soil drainage can depend on the balance of minerals and _____ matter. Clay is _____ and will not allow water to pass through so can become waterlogged. Sandy soils are _____ so water passes through too quickly.

2   In the space below draw a diagram to show how earthworms help to develop the soil profile. Use the labels to help you, and name the soil layers.

3   Give your diagram a title.

Earthworms mix soil. They rise to the surface and tunnel down.

Earthworms loosen soil. Their thousands of tunnels allow air and water into the soil.

Earthworms carry plant material to lower levels.

Earthworms eat dead plant matter and soil.

Earthworms process soil and plant materials. They deposit minerals and organic matter in casts.

Title:

2.5 How does the biosphere produce natural resources?

There is a unique atmosphere and biosphere at the Equator

1 Match the correct endings to the beginnings of these sentences.

Beginnings
The hydrosphere and atmospheric conditions at the Equator produce a climate that is
The plants use enormous quantities of carbon dioxide during
Trees grow luxurious and huge so quickly they form a dense
'Emergent' trees grow taller than the main forest canopy
They adapt to this
Other plants like lianas or vines reach the light by climbing
Large plant leaves are protected from tearing in the rain or getting fungal diseases by having
Underneath the canopy it is so dark that shade
Dead leaves create a decomposing nutrient-rich layer on top of the soil but
The unique vegetation supports and is supported by many different animals

Endings
meaning that the rainforest biome is the most biodiverse land biome in the world.
tolerant plants like mosses and ferns grow on the forest floor.
around tree trunks for support.
drip tips and waxy protective coatings.
to reach more light.
hot all year with plenty of rainfall.
photosynthesis, making the rainforest the primary global carbon store keeping the atmosphere healthy.
vegetation and rain remove nutrients, making the soil underneath poor.
canopy creating competition for light.
by having very tall, thin trunks supported by triangular-shaped buttress roots

2 Read the following paragraph.

> The Kapok tree grows up to 4 m a year. It can be 77 m high, has a thorny trunk, buttress roots 15 m across and a canopy of 61 m. Its leaves are split into deep lobes. Its flowers attract bats. Birds, frogs, spiders and monkeys live in its branches. Every part has a medicinal use – to treat asthma, fever or kidney disorders. The trunk is excellent timber for canoes and the seed threads make a stuffing for cushions and soft toys.

Tick the correct box to say whether the paragraph provides enough evidence to suggest the following statements are true:

	Enough evidence	Insufficient evidence
a) Plants in the rainforest grow quickly		
b) The Kapok tree is probably an emergent		
c) Few animals live in rainforest trees		
d) Rainforest trees are all very tall		
e) The Kapok tree provides humans with valuable resources		

4 The Kapok leaf is large and deeply lobed. Suggest one reason why the lobes might help the tree:

get more light _____

survive high wind _____

survive heavy rain _____

The hydrosphere provides an essential natural resource

1 Imagine you are a journalist. You are asked to write an informative article entitled 'At the turn of a tap' about the way your local water company provides services to ensure a clean water supply **all year.** Research your article (visit www.unitedutilities.com). You should provide four short, headed paragraphs with information about:

a) where they source the water

c) how they ensure the water is safe to drink

b) how they store the water

d) how they prevent contamination by waste water

Use the following words in your article:

drinking water	reservoir	clean water	water quality	water testing	treatment plant
sewage	water source	boreholes	micro-organisms	waste water	

_____

_____

_____

_____

_____

_____

_____

2 Research and write down three ways in which the water supply in the UK is an improvement on that in rural Ethiopia.

a) One way in which the water supply in the UK is an improvement is _____

_____

b) A second way in which the water supply in the UK is an improvement is _____

_____

c) A third way in which the water supply in the UK is an improvement is _____

_____

3 Name three important life improvements a girl in rural Ethiopia could have if a clean water standpipe were installed in her village. Use one or two words to describe each improvement.

1 _____ 2 _____ 3 _____

2.7 Why is the world so dependent on oil resources?

Oil is used for a wide range of purposes

1 Use these clues to help you complete the puzzle to reveal the key word.

1 Oil and gas are found under these.

2 Oil product used by farmers for adding nutrients to soils.

3 An oil product used every day becoming a huge environmental problem on land and in the ocean.

4 Place where crude oil is first broken down into useful components.

5 Articles we wear made from petrochemicals.

6 Power generated from burning oil as a fuel.

7 Semi-solid form of natural or refined oil used for roofs and roads, known as pitch or bitumen.

8 Another name for being reliant on oil.

9 Small particles deposited on the bottom of the ocean.

10 Transportation system for liquid oil.

11 The act of making oil into other products.

12 Vehicles using petrol as a fuel.

13 A type of fabric made from oil products.

14 Form of compressed gas made from oil.

15 Ships used to transport oil.

16 Device which powers a car using an oil product.

17 'Black gold' that comes from underground.

18 Oil, gas and oil are examples of this.

19 Areas of polluting oil at the surface of water.

20 Asphalt used on roads.

21 Unrefined form of oil originally burned in lamps in the USA.

2 On the grid, use one coloured crayon to highlight the products made from oil and another to highlight formation, extraction and processing words.

3 Write down one reason that explains why the world has become so dependent on oil as a resource.

_____

_____

_____

_____

_____

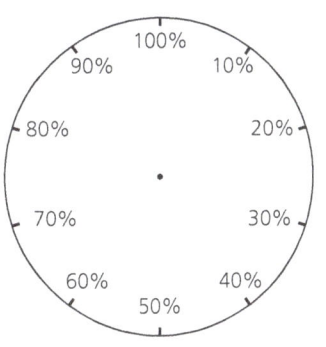

2.8 What natural resources can be used to generate electricity?

Resources for electricity generation are changing

1 Look at the pie chart A about the sources of electrical energy in the UK for 2011.

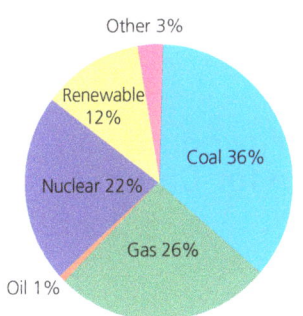

Pie chart A: Sources of electricity generation, 2011

Pie chart B

Coal	9%
Gas	42%
Oil & other	3.5%
Nuclear	21%
Renewable	24.5%

Complete pie chart B to show the share of UK electricity generation for 2016 from the table above.

2 a) Give two energy sources that increased dramatically from 2011 to 2016.

One electrical energy source that increased from 2011 to 2016 was _____

A second electrical energy source that increased from 2011 to 2016 was _____

b) Name one source which has declined dramatically.

An electrical energy source that has declined dramatically is _____

3 Study pie chart C, showing the resources China used to generate electricity in 2016, and some internet headlines.

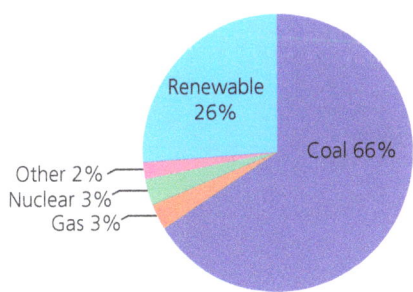

2014: Has China reached peak coal?

China fast-tracks nuclear energy

China to invest £292 billion in renewable power by 2020

Pie chart C: China's electricity generation by source, 2016

a) State one way the UK's energy mix was different to China's energy mix in 2016 and one way it was similar.

One way the energy mix in 2016 was different was _____

_____

One way the energy mix in 2016 was similar was _____

_____

b) Suggest one way you think pie chart C might change in the next few years.

_____

_____

_____

_____

2.9 How can we use natural resources sustainably?

1 Study the headlines about ways to reduce the use of electricity produced from fossil fuels in the UK and the world. Decide if they are actions that could be made on an individual, a local, a national or an international scale.

Choose one colour crayon for each scale of category and shade the headlines in the correct colour. Add a key. If the action operates at more than one level, use stripes.

Government announces grants for cavity wall insulation schemes Council must put into operation

New energy education campaign for Cheshire schools

Application of the Paris Agreement 2020. Will president renege?

Efficient Energy Research. Government to invest in latest advances

Switch off those lights! Save money now.

Town Hall to be insulated. Council taxes will rise

EU emissions agreement made How it affects your energy bill.

Key		Individual		Local		National		International

2 **a)** Think about your school. Make two suggestions that would make it more sustainable by reducing energy consumption or waste. Explain how each suggestion would help.

Suggestion 1_____   This would help by _____

_____   _____

Suggestion 2_____   This would help by _____

_____   _____

b) Think of one thing you could do at school to save energy.

Suggestion 3_____

This would help save energy by _____

_____

_____

_____

_____

2.10 How do we use our planet as a natural resource? Review

2.10

Student's Book
page 40–41

You have been asked to produce a poster for the United Nations explaining how the Earth's systems work. Use your plan and the reminders below to produce a final version using IT. Stick your poster in the space below over the reminders.

Stick your poster here.

Remember to consider:

the name of the organisation

the title of the poster

your background colour

the font and size of the wording

the colouring of the words and pictures/diagrams

how your pictures will represent resources and Earth systems and their interactions

which resources you have selected to depict

which facts you have decided to depict

how you will show the facts you think are most important

3.1 What is an economy, from local to global?

Job structure in the UK has changed over time

1 Think of a job from each of the four sectors you have learned about:

Primary sector job: _____   Tertiary sector job: _____

Secondary sector job: _____   Quaternary sector job: _____

2 For each job, create an advert to place in a newspaper. Include the job title, the kind of work the applicant would be expected to do, where that work would take place and the skills he or she should have.

Primary sector	Secondary sector

Tertiary sector	Quaternary sector

3 Study Graph A, which shows the change in employment structure in the UK since 1791, and the labels around it. Make it into a 'living graph' by drawing arrows from the labels to the points on the graph that they explain.

> UK factory workers make goods for the whole world

> Farmers employ a huge workforce especially at harvest

> Wages are low; people cannot afford holidays or doctors

> In the consumer age, people have disposable income with which to eat out, buy goods and enjoy entertainment

Graph A: Economic sectors in the UK from 1791 to 2011

> UK factories decline in the face of cheap imports

> Modern farms use labour-saving machinery and computers

> New medicines and computer programs are researched

3 a) Suggest a change that you think may take place in one employment sector in the next 50 years. _____

b) Give one reason for your answer. _____

© Hodder Education 2018

3.2 What's happening down on the farm?

1 Look at the statements below. Are they true or false? Correct any that are false.

	TRUE OR FALSE
Farming is in the secondary economic sector.	
Arable farms grow crops.	
Pastoral farms have crops and animals.	
A market garden raises only animals.	
Farms that are run as businesses are commercial.	
Farms that are run only to feed the farmer and his family are called mixed.	
Market gardens and dairy farms need good transport systems because their products are perishable (do not keep well).	

2 Not all market gardens are small. Use the internet to research Thanet Earth, Kent, the largest glasshouse complex in the UK – www.thanetearth.com. Complete the system diagram below for the Thanet Earth complex using the information you have gathered. Add at least four inputs, three processes and two more outputs.

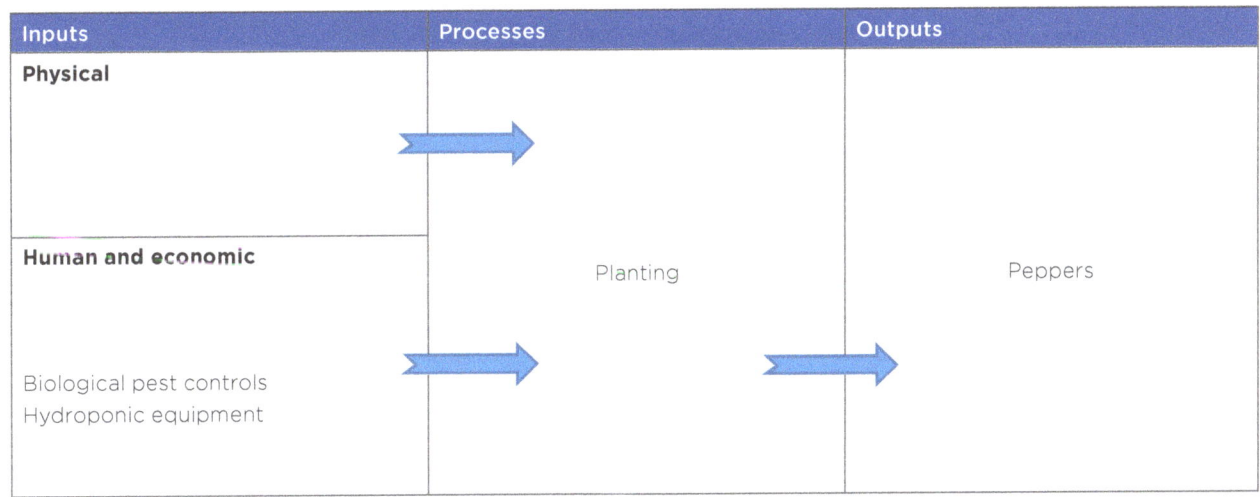

Inputs	Processes	Outputs
Physical		
Human and economic	Planting	Peppers
Biological pest controls Hydroponic equipment		

3 Once, market gardens had to be close to cities and have warm climates. Explain how the following factors have helped farmers create larger, intensive farms further from their markets.

Greenhouse technology _____

_____

Better roads and refrigerated lorries _____

_____

3.3 Why is manufacturing all about choosing the right site?

The importance of factory location

1 Read the following talk given by a business owner looking to build a factory. She is describing to other members of her team a location she has found for her new factory.

   Complete the talk by filling in the blanks with words from the box below.

people	network	markets	expensive	skills	raw
	Europe	materials	cost	finished	government

'This factory site has a lot of positives for us! We could employ local _____ from the large town nearby; many have the _____ and experience for our type of work already.

The road _____ is good , so we can get _____ _____ in and our _____ product out. We are close to our _____ in the UK and in _____.

The _____ would have to support us if we choose to build here because the _____ of buying this land for our factory is very _____.'

2 a) Identify one advantage of this location for a new factory_____

   _____

   b) Identify one disadvantage of this location for a new factory_____

   _____

3 a) Would you locate your factory here? **YES / NO**

   b) Explain why you have made this choice.

   _____

   _____

4 The five 'Cs' used to find the right location for a factory are: **Costs, Capacity of workforce, Capabilities of the region, Culture of the region, Customers.**

   a) Arrange each of the five 'Cs' in the diagram in the order you think is most important. Start with the most important in the top box.

   b) Explain the reasons for your top and your bottom choice.

Most important choice: _____

_____

_____

Least important choice: _____

_____

_____

Most important

Least important

3.4 Why did Nissan locate in the UK?

Location choice can give an industry advantages

1 Read the statements about the UK Nissan plant in Sunderland.

Draw lines from the statements to complete the timeline.

1951
14,000 employed
in shipbuilding

1984
Government
incentives
support Nissan
purchase of
cheap land

1986
500 employed;
5,000 cars
produced

1988
Nissan
employs
1,300

1994
Exports
to 36
countries

2012
Over 510,000
vehicles
produced
each year

2014
Nissan
announce
skills foundation
and school link

1951	1961	1971	1981	1991	2001	2011	2021

1950s
Slump in coal
use, pits close

1984
Only one
shipyard
left with 4,000
workers

1986
PM Margaret
Thatcher opens
the new plant

1988
Last
shipyard
closes

2000
Over 300,000
vehicles
produced;
now Britain's
leading car
producer

2012
Export to
130 countries
through port

2016
Over 7,000
employees

Timeline of Nissan development

2 Suggest three reasons why Nissan's choice of Sunderland for their factory made in 1984 was a good one.

Reason 1: _____

Reason 2: _____

Reason 3: _____

3 Examine the pie chart and map below.

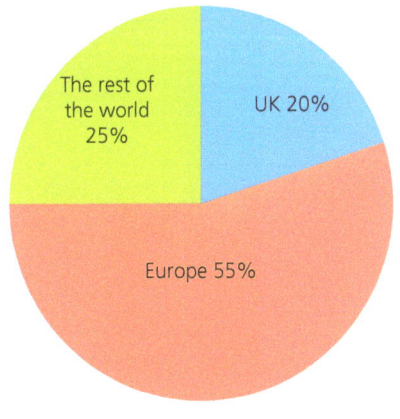

The rest of
the world
25%

UK 20%

Europe 55%

Pie chart A: Nissan's markets

Sunderland

Europe

N

0   500 km

Map A: Sunderland's position relative to Europe

a) What percentage of Nissan's output is sold in the UK? _____ %

b) What percentage of its output does it sell to Europe? _____ %

c) What percentage does it sell to the rest of the world? _____ %

4 What benefit does Sunderland's location have for Nissan to sell its products in Europe?

_____

3.5 Why is the tertiary sector increasing?

Tourism in the UK is an expanding industry

1 Number these events in the correct chronological order to help explain UK tourist development. Give the earliest event number 1 and the most recent event number 8.

The railways connected people to holiday resorts cheaply and quickly.

Increased pay and bank holidays allowed people to make day trips.

Cheap air travel took UK holidaymakers abroad.

People began to be able to own cars, which together with more leisure time meant more UK holidays.

Tourism began in the UK in about 1850.

The UK became a popular destination for people from abroad with historical buildings, famous film locations and designer shopping for wealthy Asians.

Workers' wages hardly bought enough food or paid the rent. Only the very wealthy could afford holidays.

A record 41 million visitors are a major boost to the UK economy.

2 Study the list of occupations associated with tourism in the box below.

Café owner   Brontë lecturer   donkey ride provider   waitress   mountain bike hirer
beach lifeguard   camping gear provider   hotelier   museum guide   walking boot seller
Brontë souvenir seller   oyster seller   boat hire provider   book shop owner

a) Place each job where you think it is most likely to take place in the Venn diagram below.

Yorkshire coast

North York Moors

Haworth (Brontë) village area

b) Name three jobs you think would provide income in any tourist area.

i _____

ii _____

iii _____

c) Choose one of the jobs and explain why it would provide income in any tourist area.

_____

_____

3.6 How does a chocolate bar connect the sectors of the economy?

3.6

1 Complete the sentences using the words from the box.

> Britain   chocolate bars   container ship   beans   Ghana   lorry   factory   ship

The raw material cocoa pods are grown in countries like _____ and Ivory Coast.

The dried cocoa _____ are sent by _____ to countries like _____.

The beans and other ingredients are made into _____ in a _____.

Bars are exported by _____ to other countries.

The bars are sent by _____ to the shops and supermarkets.

2 Read the speech bubbles below. Which type of job fits with each speech bubble – a primary, secondary, tertiary or quaternary job?

> I sell chocolates in my shop.

> We find out which gases make the best bubbles in chocolate.

> I grow cocoa on my small farm.

> I make chocolate in the factory.

3 Chocolate travels a long way to reach you, the consumer:

- Most Ghanaian cocoa is transported via the port at Tema eventually to Liverpool – over 7,000 km by sea

- Nestlé cocoa is transported from Liverpool to York – 160 km

- Cadbury's cocoa is transported from Liverpool to Chirk – 78 km – for roasting and pressing

- From Chirk to Leominster it travels 99 km to have local milk and then sugar added

- From Leominster to Bournville it travels 85 km to produce the final product

- From Bournville it travels to your local shops or on for export to the EU

Use an atlas map of the UK to help you plot on the map the journey of a Cadbury's chocolate bar from Liverpool to your local shop.

Map A: Chocolate's UK journey

3.7 How does the UK trade with other countries?

1 Read the following sentences and complete them by circling the correct word.

a) Things a country buys in from abroad are called **exports / imports / ports**.

b) Things that a country sells to another country are called **exports / imports / ports**.

c) The difference between the money made from imports and exports is called the **balance of trade / imbalance of trade / trade**.

d) Imports and exports include **goods / services / goods and services**.

2 a) Use the figures in the table to complete the dual bar chart for February 2018 UK trade with leading partners. Colour imports red and exports blue.

b) Label the vertical axis 'Trade in billion £' and the horizontal axis 'Country'.

c) Give the graph a title.

UK trade with leading partners in February 2018	
Imports	**Exports**
Germany £5.8 billion	Germany £3.0 billion
Netherlands £3.2 billion	Netherlands £2.0 billion
USA £3.0 billion	USA £3.6 billion
France £2.2 billion	France £1.9 billion

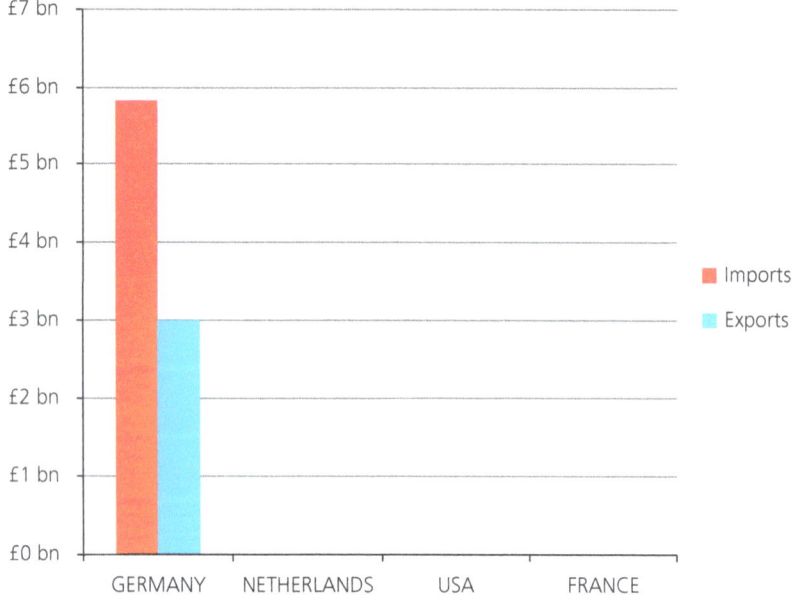

3 a) Which country bought more things from the UK than it sold them in February 2018?

_____

b) Which country exported the most to the UK in February 2018? _____

c) What was the total value of imports to the UK from these four countries for the month?

£ _____

d) What was the total value of exports from the UK to these countries for the month?

£ _____

e) Did the UK make a profit in its trade or a loss? _____

3.8 What is globalisation?

Globalisation affects most aspects of UK life every day

1 Investigate the contents of your fridge and kitchen cupboards. List 10 products and their origins.

Food/kitchen item	Country of origin	Food/kitchen item	Country of origin

2 Construct a bar chart or a pie chart to help visualise your results.

_____ chart to show the origin of food in my kitchen

3 Write one sentence to describe what the sample contents of your cupboards/fridge show about globalisation.

_____

_____

3.9

3.9 How has containerisation accelerated globalisation?

Container ships transport goods around the world

1 Log onto the marine traffic website: www.marinetraffic.com and click 'vessels'. Type 'OOCL HONG KONG Container Ship [HK]' into the search box to locate the ship.

2 Complete the log by recording its position (by latitude and longitude and the nearest country or sea area it is passing) each day for a week. Work out which of the following ports it is travelling to and which it called at last:

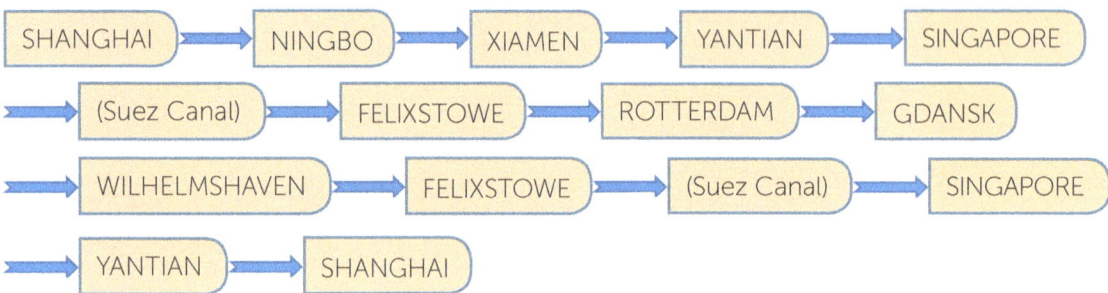

SHANGHAI → NINGBO → XIAMEN → YANTIAN → SINGAPORE →
(Suez Canal) → FELIXSTOWE → ROTTERDAM → GDANSK →
WILHELMSHAVEN → FELIXSTOWE → (Suez Canal) → SINGAPORE →
YANTIAN → SHANGHAI

3 Mark the ports on the map and plot the ship's position each day.

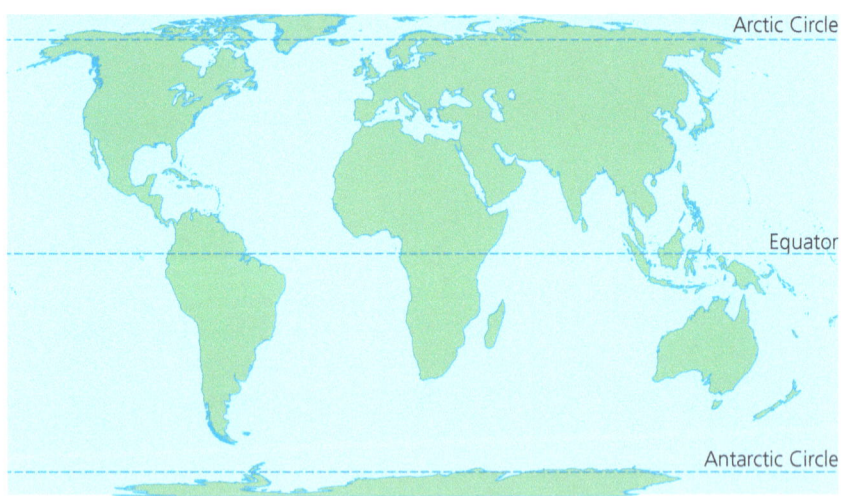

Map A: The route of the OOCL *Hong Kong*

	Day 1	Day 2	Day 3	Day 4	Day 5	Day 6	Day 7
Date							
Time							
Last port visited							
Current location (nearest country/sea area)							
Latitude and longitude							

4 Click on 'Companies at destination'. Suggest two types of goods the ship is carrying.

3.10 What is an economy, from local to global? Review

1 Use the clues about economy to complete the crossword.

Clues across	Clues down
2 Movement of goods from one location to another.	**1** The money made when trading.
4 and 3 down A Chinese factory making iPhones.	**2** Buying and selling goods.
5 A farm that only grows crops.	**3** See **4 across**.
8 A huge cargo-carrying box for a ship or lorry.	**6** When the cost of imports is more than the cost of exports.
11 A car company in Sunderland.	**7** A good that leaves the country.
12 Jobs such as farming, mining and fishing.	**9** Jobs providing a service.
16 Vessels that transport goods across the sea.	**10** Farms rearing animals only.
18 Farms with crops and animals.	**13** Goods brought into a country.
19 Large companies with offices and factories worldwide.	**14** Country that exports cocoa beans.
20 One of the 'Cs' that companies have to consider when locating a factory.	**15** Items made or sold.
23 Flows of people, goods and ideas worldwide.	**17** A job sector where things are made.
24 Jobs involving research and development.	**21** A place where ships bring cargo into a country.
	22 The path a ship takes between certain ports.

4.1 What is weather and climate?

1 Use the clues about the weather to complete the crossword.

Clues across	Clues down
1 The layer of gases surrounding the earth.	**2** A form of precipitation in cold weather.
3 Too much rain may cause one of these.	**4** Common in a thunderstorm.
5 Too much rain may cause crops to do this.	**6** A measure of how hot or cold it is.
7 A long period of dry weather.	**8** A common type of precipitation.
9 The description of the weather when it is raining.	**10** ---centres are found where there is frequent snow.
13 Rain, hail, snow and drizzle are all examples of this.	**11** This activity is made difficult by foggy weather.
14 Average weather conditions.	**12** Common in winter.
16 See **15 down**.	**15 and 16 across** Force exerted by the weight of the air, measured in millibars.
19 Suspension of water droplets causing visibility less than 1000m.	**17** This is good for crops but too much can shrivel them.
22 The amount of water vapour in the atmosphere.	**18** A mixture of snow and rain.
23 The day to day changes in the atmosphere.	**20** A suspension of water droplets with visibility above 1000m.
24 and 21 down A type of severe weather.	**21** See **24 across.**
25 A type of precipitation with very small droplets of water.	**23** The movement of air from an area of high to low pressure.

4.2 How do we measure weather?

4.2

Recording weather measurements

1 Use the table to record the weather readings you take for the period of a week from your school weather station. Take the readings at the same time each day.

If your school has no weather station, use the internet to find the readings for your local area. Take care to notice how they measure wind speed.

Weather record for the week:							
Season:							
Location:							
DAY	1:	2:	3:	4:	5:	6:	7:
TIME of measurement							
Temperature maximum	°C	°C	°C	°C	°C	°C	°C
Temperature minimum	°C	°C	°C	°C	°C	°C	°C
Rainfall	mm	mm	mm	mm	mm	mm	mm
Air pressure	mb	mb	mb	mb	mb	mb	mb
Sunshine	hours	hours	hours	hours	hours	hours	hours
Cloud (high/medium/low)							
Wind speed (1 knot = 1.852 km/h)	knots	knots	knots	knots	knots	knots	knots
Wind direction							
Notable weather event							

2 Complete the following paragraph to describe the week's weather using the data from the table.

The weather in the study week was generally _____

_____ .

The highest daytime temperature was _____ on _____ .
The lowest daytime maximum temperature was _____ on
_____. The minimum night time temperature was _____
on _____. Sunshine levels and duration were _____
and cloud _____. The wettest day was _____ when
_____ mm equivalent of rain fell. The driest day was _____
when _____ mm of rain fell. A total of _____ mm of rain
equivalent fell during the week. Wind blew mainly from the _____ and
reached speeds of _____. The most notable feature of the weather was
_____. It occurred on _____ at _____
hundred hours.

4.3 How can weather data be recorded and presented?

Using the synoptic code to present weather data

1 Find a weather map or chart for your local area from a newspaper or website (or type 'local weather' into a search engine) and stick it in the space below.

2 **a)** Choose one place on your map and mark it A.

   b) Use the information from the weather map to complete a synoptic diagram for point A.

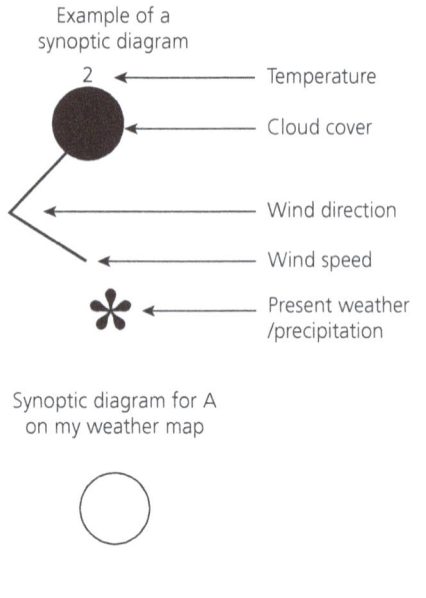

Example of a synoptic diagram

2 ← Temperature

← Cloud cover

← Wind direction

← Wind speed

← Present weather /precipitation

Synoptic diagram for A on my weather map

Symbol	Precipitation	Symbol	Cloud cover	Symbol	Wind speed
ҙ	Drizzle	◯	Clear sky	◎	Calm
▽	Shower	◐	One oktas	◯—	1–2 knots
●	Rain	◔	Two oktas	◯⌐	5 knots
✳	Snow	◕	Three oktas	◯⌐	10 knots
△	Hail	◑	Four oktas	◯⌐	15 knots
ꓘ	Thunderstorm	◒	Five oktas	◯⌐	20 knots
⸪	Heavy rain	◕	Six oktas	◯▼	50 knots or more
⁎	Sleet	◑	Seven oktas		
✳▽	Snow shower	●	Eight oktas		
═	Mist	⊗	Sky obscured		
≡	Fog		The sky is divided into eighths or oktas to record how much cloud cover there is.		

Standard symbols used on weather charts

3 Write no more than two sentences to describe the weather shown on the synoptic chart/ weather map._____

_____

The different types of rain affecting the UK

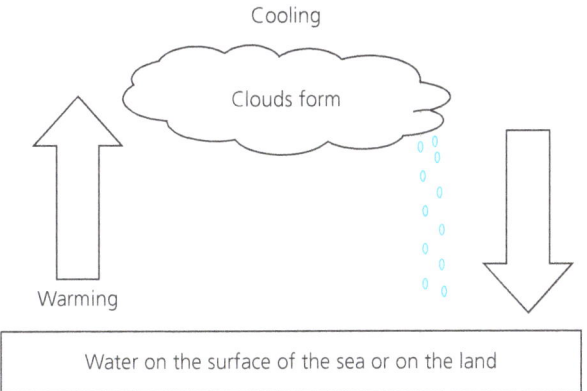

Cooling

Clouds form

Warming

Water on the surface of the sea or on the land

The basic rain cycle

1 Add the three labels in the correct places on the rain cycle diagram.

Water evaporates to water vapour

Water vapour condenses to water droplets

Precipitation as rain

2 Look at the three maps showing the different types of rainfall in the UK. Using your knowledge of the different types of rain, complete the table in your own words.

Key Uplands Rainfall intensity 8 mm per hour 6 mm per hour 4 mm per hour 2 mm per hour 0 mm per hour 0 N 200 km	14.00 hours 2 April 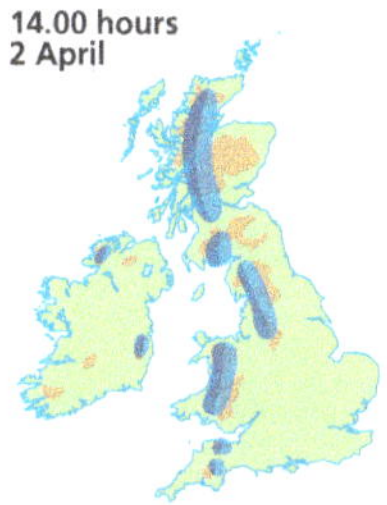	09.00 hours 21 November 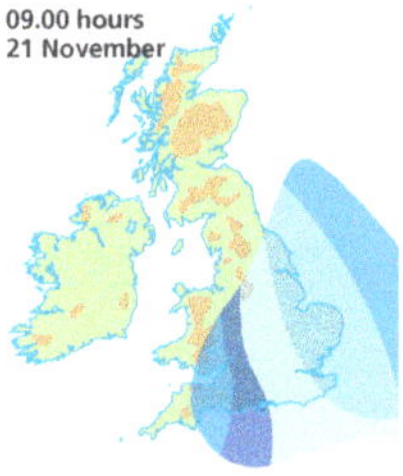	14.00 hours 9 June 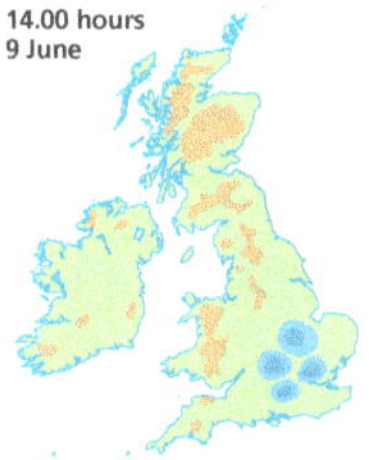
Cause of evaporation	Warm westerly winds blow over the sea		Intensive daytime heating of the land
Cause of uplift and cooling		Cold dense air moves faster and lifts the warm moist air off the ground	
Typical rain character			Intensive rain with thunder and lightning
Typical location and area	Falls to the west and tops of mountains The east is a rain shadow or drier area		
Name of rainfall		Frontal	
Usual time of year this rain falls		All year, notably winter	

4.5 What are air pressure and anticyclones?

Contrasting winter and summer anticyclones

1 Read the following real impacts (effects) of anticyclones that have affected the UK in the recent past.

Older people advised to keep one room warm

35°C temperatures wilt crops

Burst pipes close schools

Intercity train disruption: rails buckle

The entire stock of a Liverpool chocolate factory melts

Major blackouts: power cables snap!

Elderly residents evacuated from New Forest care home as forest fire approaches

M6 chaos: freezing fog pile-ups

Fracture queues: NHS under ice strain

2 Highlight the impacts of a winter anticyclone in blue and those of a summer anticyclone in red.

3 Choose one of the impacts and write a short newspaper report describing the event. Give your report a sensational headline.

Don't forget to say:

- when it happened (summer/winter)
- where it happened
- who was affected
- how they were affected
- why it happened.

You could include a short interview with a person affected by the weather and a photograph or drawing.

4.6 What are depressions and how do they affect our weather?

4.6

Student's Book
pages 72-73

Weather forecast for a UK depression

1 Study the script for this morning's school weather forecast. The forecaster will use Map A to help the presentation.

Circle the correct words to complete the script correctly.

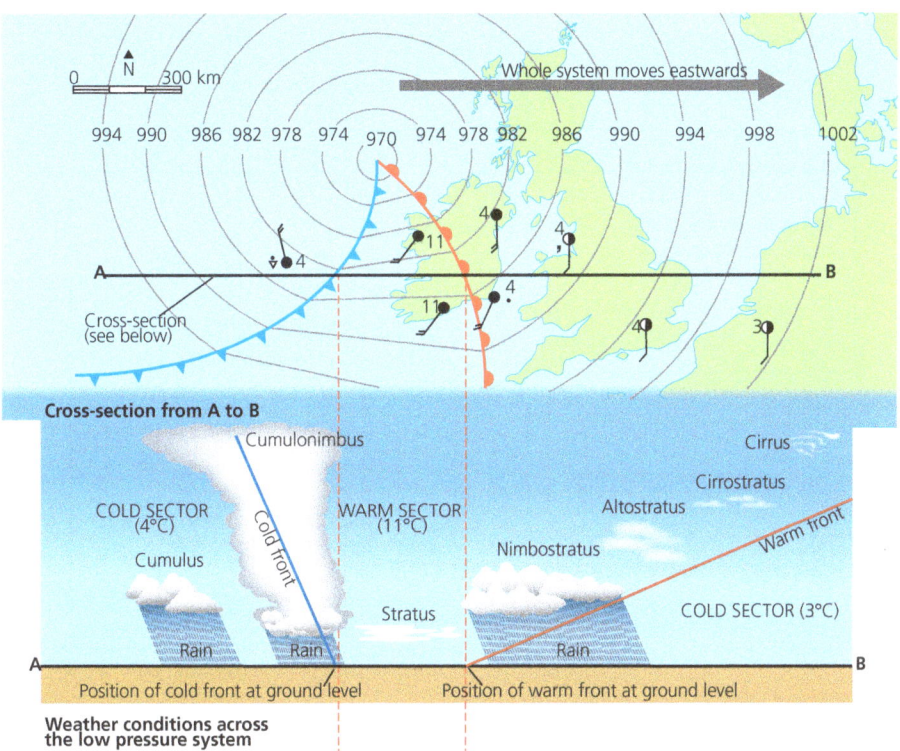

Map A: UK weather map

Good morning.

It's going to be a **chilly / warm** walk to school at **11°C / 4°C** so take care.

Pressures are **high / low** and set to **fall / rise** more so – wherever you are in England or Wales, it's almost certain you will **remain dry / experience rain** today. Mums will say, 'Take a **sunhat / waterproof.'**

Sun / Drizzle has already reached areas on the **east / west** coast.

Don't let those **high / low** clouds fool you. They will lower and thicken giving **drizzle / sunshine** followed by rain by breaktime. However, temperatures will rise sharply to **double / single** figures once the warm front passes you.

Rain may ease or even clear by lunchtime, but make the most of it – the rain may not stop for long, if at all. That **cold / warm** front will catch up fast. The field may be **usable / out of bounds** again!

The front will definitely bring **heavy / light** rain showers from cumulonimbus clouds by **afternoon games / home time**, followed by a change in the wind direction. A soggy end to the school day.

This wind will blow from the **north / south** or **south-west / north-west** bringing those cold, cold temperatures again. The rain will slow to a drizzle before clearing away but if you're already wet the wind will feel really cold. If you forgot your coat, Mums will comment 'I told you so.'

There is another area of low pressure sitting to the **east / west** in the Atlantic, waiting to pass tomorrow, so another **umbrella / warm coat** day to follow! Take care today.

4.7 How do I conduct a weather enquiry?

The relationship between weather systems and weather

1 Study the mind map for weather associated with anticyclones and add anything you think is missing.

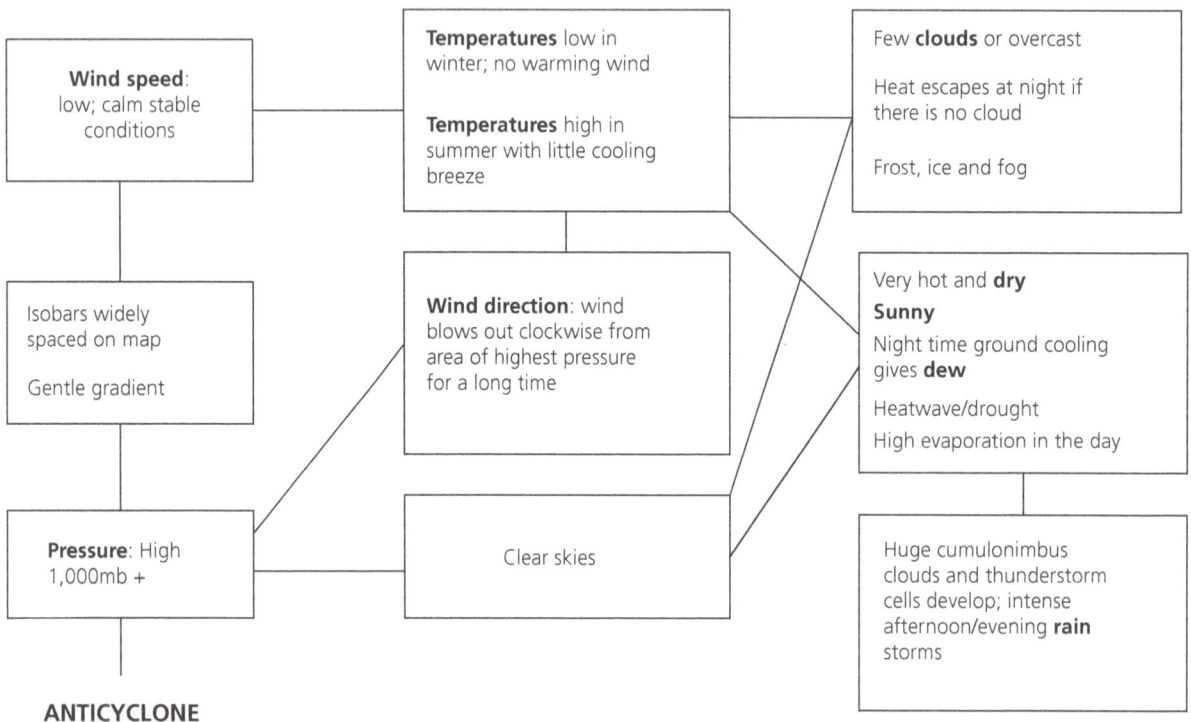

Wind speed: low; calm stable conditions	**Temperatures** low in winter; no warming wind **Temperatures** high in summer with little cooling breeze	Few **clouds** or overcast Heat escapes at night if there is no cloud Frost, ice and fog
Isobars widely spaced on map Gentle gradient	**Wind direction**: wind blows out clockwise from area of highest pressure for a long time	Very hot and **dry** **Sunny** Night time ground cooling gives **dew** Heatwave/drought High evaporation in the day
Pressure: High 1,000mb +	Clear skies	Huge cumulonimbus clouds and thunderstorm cells develop; intense afternoon/evening **rain** storms

ANTICYCLONE

Anticyclone mind map

2 On a piece of paper create a mind map for depressions and make as many connections as you can. Use the map and exercises from pages 4.5 and 4.6 on anticyclones and depressions, and Map A below. Stick your mind map on one side to create an overlay so you can compare them both.

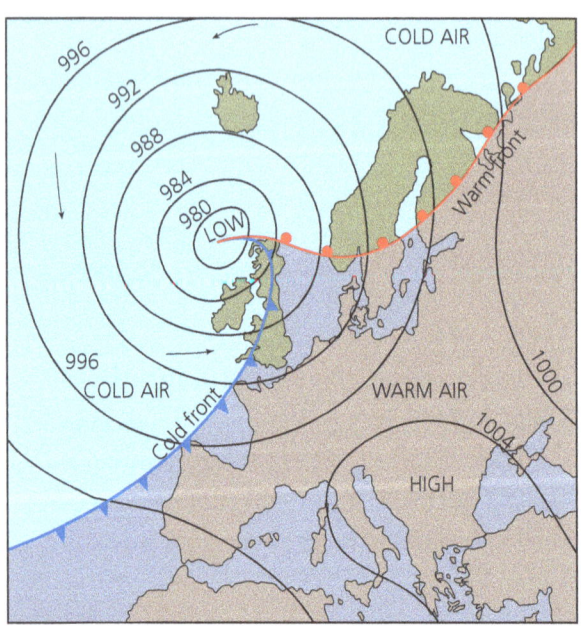

Map A: Synoptic map of high and low pressure weather

4.8 What is the climate of the UK?

How the climate of the UK affects occupations

1 Complete the following statements by completing the blanks.

Climate is the _____ conditions of weather calculated over _____ years.

Weather is the minute _____ _____, day to day _____ in weather conditions.

2 Study the map below, which divides the UK into four climatic regions.

Hill sheep farmer
_____
_____
_____
_____
_____
_____

Ski resort owner
_____
_____
_____
_____
_____
_____
_____

North West
Winters cold
Jan. 3°C
Summers mild
Jul. 13–14°C
Pptn<750 mm

North East
Winters cold
Jan. 3°C
Summers mild
Jul. 13–14°C
Pptn
600–900 mm

South West
Winters mild
Jan. 7°C
Summers warm
Jul. 15–20°C
Pptn>1000 mm

South East
Winters cold
Jan. 3°C
Summers warm
Jul. 15–20°C
Pptn<500 mm

Seaside hotelier
_____
_____
_____
_____
_____
_____
_____

Crop/arable farmer
_____
_____
_____
_____
_____
_____
_____

Describe at least two ways the climate of each of the areas helps the jobs these people have. Give one difficulty each climate might present for them or for other activities in their area.

3 a) Where in the UK do you live? Write two sentences to describe your local climate.

_____

_____

_____

b) Name one economic activity in your area affected by the climate.

_____

_____

4.9 How does climate vary across the world?

Factors that affect temperature across the world

1 Complete the sentences by matching the beginnings to the correct endings.

a) A small area of land at the Equator receives	i) every 100m increase in altitude
b) Other factors like altitude, prevailing winds,	ii) and can transfer warmth from warm ocean currents onto the land.
c) Temperatures decrease by 1°C for	iii) ocean currents and distance from the sea also influence temperature.
d) Prevailing winds blow from one direction for most of the time	iv) but it cools down much more quickly in the winter.
e) Land heats up much more quickly than the sea in summer	v) the same amount of solar heat as a larger area nearer the poles.

2 Read through your completed sentences carefully and match each sentence a–e to **two** of the following 'This means that' statements below.

THIS MEANS THAT ...

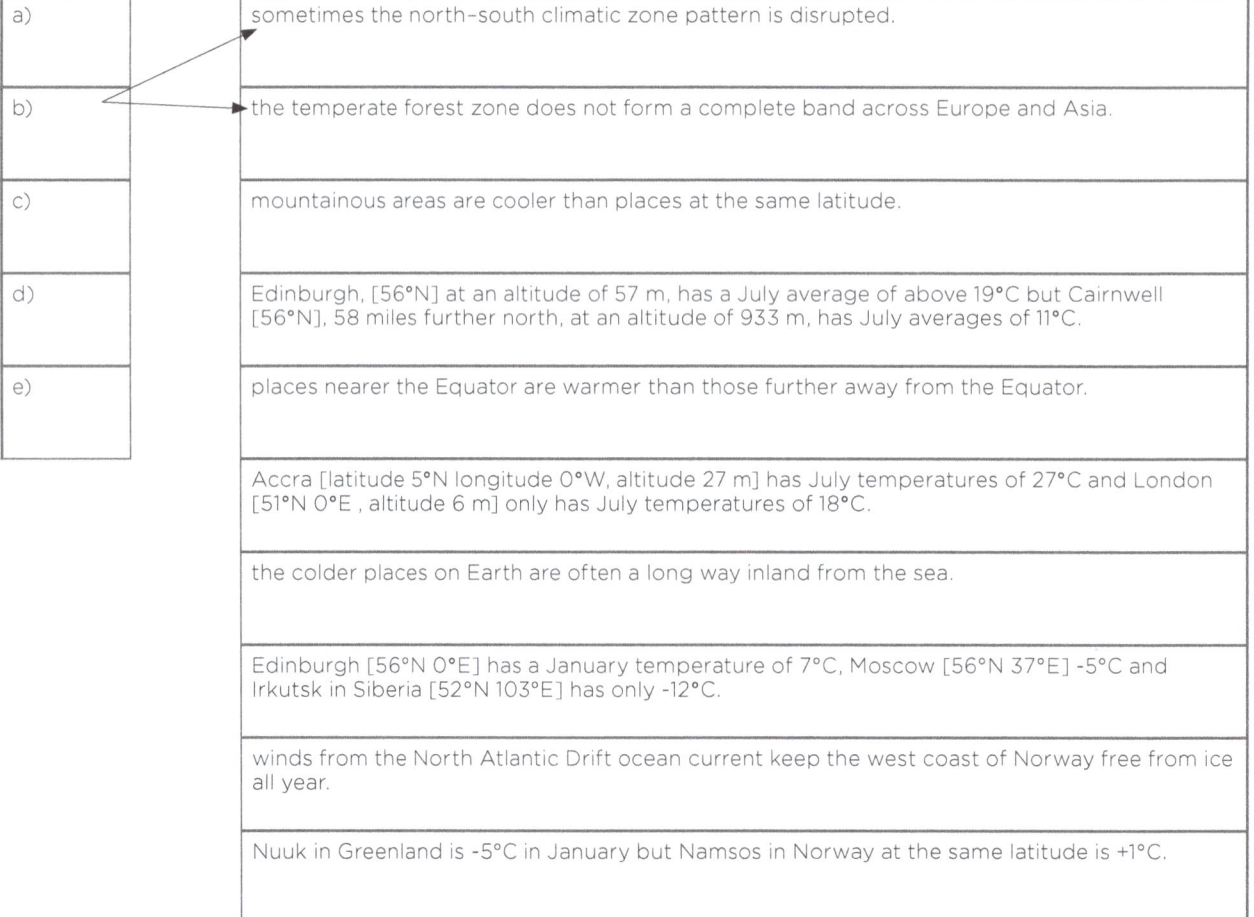

a)	sometimes the north–south climatic zone pattern is disrupted.
b)	the temperate forest zone does not form a complete band across Europe and Asia.
c)	mountainous areas are cooler than places at the same latitude.
d)	Edinburgh, [56°N] at an altitude of 57 m, has a July average of above 19°C but Cairnwell [56°N], 58 miles further north, at an altitude of 933 m, has July averages of 11°C.
e)	places nearer the Equator are warmer than those further away from the Equator.
	Accra [latitude 5°N longitude 0°W, altitude 27 m] has July temperatures of 27°C and London [51°N 0°E , altitude 6 m] only has July temperatures of 18°C.
	the colder places on Earth are often a long way inland from the sea.
	Edinburgh [56°N 0°E] has a January temperature of 7°C, Moscow [56°N 37°E] -5°C and Irkutsk in Siberia [52°N 103°E] has only -12°C.
	winds from the North Atlantic Drift ocean current keep the west coast of Norway free from ice all year.
	Nuuk in Greenland is -5°C in January but Namsos in Norway at the same latitude is +1°C.

4.10 What is weather and climate? Review

1 Use the clues to complete the word puzzle and reveal a type of atmospheric condition.

1 Body of uniform air

2 Type of raincloud

3 Instrument to measure wind speed

4 Type of rain caused by moist air rising over hills

5 Direction of wind blowing into an area of low pressure is anti-_____

6 A measure of moisture in the air

7 Frozen water

8 Average weather conditions over time

9 Period without rain

10 Rainfall associated with thunder and lightning

11 Another word for height

12 Type of weather map using a special code to represent weather station observations

13 Type of air pressure associated with the centre of a depression

14 Weather system associated with fronts

15 Areas where warm and cold air masses meet

16 Measure of distance from the Equator

17 Pressure associated with an anticyclone

18 Instrument used to measure temperature

19 A blue line with triangles on a weather map

2 Name two features of weather you would associate with the clueword answer.

_____ and _____

Locating features in Russia

1 On Map A, 'The location of Russia', label the following lines of latitude: the Arctic Circle – 66½°N and 50°N and the following lines of longitude: 170°E, 170°W, 30°E and 80°E.

Map A: The location of Russia

2 Use an atlas map to help say if the following statements are TRUE or FALSE.

Statement	TRUE OR FALSE
Most of Russia is between 20° and 40° north of the Equator.	False
Russia stretches from Kalingrad (54° North 19° East) in the west to just beyond Cape Dezhnev (66° North 169° West) in the east.	
The USA is next to Russia in the east.	
All of Russia is south of the Arctic Circle.	
The Franz Joseph Land Islands are less than 10° from the North Pole	
Russia is mostly in the western hemisphere.	
Russia is mostly north of 50° North of the Equator.	
Russia's most southerly point is roughly 41° North 47° East.	
Volgograd has a latitude of 100° East.	

3 Choose two cities in Russia. Name each city and write its approximate latitude and longitude next to it.

City 1: _____

City 2: _____

5.2 What is the physical landscape of Russia?

The distribution of physical landforms in Russia

1 Use the clues about Russia's landscape to complete the crossword.

Clues across	Clues down
2 The highest point in Russia, rising to 15,462 m; an extinct volcano: Mt _____.	**1** An ocean that forms the eastern coastline of Russia.
6 and 4 down A vast area of salt water to the north of Russia.	**3 and 14** A warm sea area with access to the Mediterranean.
9 and 17 A very volcanic part of Russia's far east projecting into the Pacific Ocean with 70 volcanoes.	**5** A large Russian occupied island north of Japan which experiences many earthquakes.
11 and 4 and 7 The Pacific _____ _____ _____, an area of earthquake and volcanic activity.	**8** The Siberian Rivers Ob and Yenisei flow north into this sea.
12 Mountain range near the Caspian Sea.	**10 and 5 across** A large inland sea, below sea level.
15 A peninsula in the European north of Russia on the Arctic coastline.	**13** Mountain range dividing European Russia from Asian Russia.
16 and 1 A huge expanse of lowland to the east of the Ural Mountains.	

5.3 What is the climate of Russia?

Factors influencing the Russian climate

Map A: Average July temperatures

Map B: Average January temperatures

1 a) Look at map A. Write one sentence to describe the general pattern of temperatures in Russia in July._____

_____

b) What factor causes this distribution of temperature? Highlight the correct factor in the boxes below.

| LATITUDE | ALTITUDE | DISTANCE FROM THE SEA |

c) Area A, near the Atlas Mountains, is colder than places the same distance north. Use another word or phrase from the boxes above to write a sentence to explain this.

_____

_____

d) Area B, Yakutsk, is much warmer than might be expected. Explain why this is the case.

_____

_____

_____

2 a) Look at map B. Write a sentence to describe the distribution of temperature in winter.

_____

_____

b) What do you consider to be the one most influential factor governing January temperatures? Justify (give reasons for) your answer. _____

_____

Plants and animals adapt to different biomes

All biomes have both plant and animal populations as part of their ecosystem which have adapted to survive in their unique environments.

Study the fact files about the Arctic fox and the red fox below. The red fox is part of the ecosystem in the Taiga and the Arctic fox lives in the Tundra. Choose two adaptations that each of the foxes has to help them survive in their biome. For each one, explain how that adaptation helps.

Tundra biome	The Arctic fox	Adaptation 1:
Climate: Winters: dark and cold, below -40°C. Summers: cool, too short for trees to grow. Precipitation: very low.	• Dense, thick, multi-layered coat • White fur in the winter • Small; compact body shape and short legs • Rounded muzzle, small ears • Only mammals with furry footpads • Live in underground burrows which may be over 100 years old • Jumping and pouncing hunting attitude • Amazing sense of smell and hearing	_____ Helps because _____ _____ _____ Adaptation 2: _____ Helps because _____ _____ _____
Vegetation: reindeer mosses, liverworts and thin grasses		
Prey animals: Lemmings, seabirds, birds' eggs, fish, dead seals		
Predator animals: Wolves, eagles, lynx		
Taiga biome	The red fox	Adaptation 1:
Climate: Winters: cold, sub-zero temperatures. Summers: may be very warm, long enough for trees to grow. Precipitation: higher than in the Tundra.	• Short, dense, warm fur • Slender body shape • Long, slender legs, good for running • Red rusty-coloured fur with white tipped tail • Pointed ears and long nose allow evaporation of heat • Lots of stamina • Good night vision	_____ This helps because _____ _____ _____ Adaptation 2: _____ This helps because _____ _____ _____
Vegetation: Coniferous trees and larches with ferns, ground plants and berries		
Prey animals: Wide variety of small mammals, worms, invertebrates		
Predator animals: Lynx, wolves		

5.5 Where do people live in Russia?

Factors affecting population distribution in Russia

1 a) In the boxes below draw the symbol of one stick man per person to show that the population density of Russia is 8.5 people/km² but in Siberia it is only 2.5 people/km².

Population density Russia	Population density Siberia	Population density European Russia

2 a) The population of European Russia is roughly 110 million people and its area is 4 million km².

Use this calculation to find its population density:

$$\frac{\text{Number of people}}{\text{Area in km}^2} = \frac{110}{4} = \text{population density} = \underline{\hspace{2cm}} \text{ people per km}^2$$

Complete the population density box with stick men symbols.

2 b) What does this show about the distribution of people in Russia today? _____

_____

3 Read the interviews with Siberian people. List five things that may attract them to Moscow or another Russian city.

Difficulties that are making people unhappy with their Siberian town or village (and may drive them to leave).	Things that attract people to live in cities in the west like Moscow or Southern Siberia. (These may draw people towards Moscow.)
Viktor, middle-aged man: We're lucky if the mobile shop arrives once a week. My wooden house is cold. There is no piped water, just wells.	Reason 1:
Anatol, retired coalminer: This town used to be so busy. Now the coal is too expensive to mine. My son's job is not secure.	Reason 2:
Alexei, 16: There are only old people here. We have no neighbours. The roads are awful. There's nothing to do. There is no college for hundreds of miles. I don't want to do farm work.	Reason 3:
Annushka, young mother: We work all day. I still have to look after my mother afterwards. My two children are young, but the school and hospital have closed. We get power cuts.	Reason 4:
Misha, farmer 54: Work is hard – we do a lot by hand. I was born here, I'm used to it. We use a cart and horse. We cannot get a tractor repair man anymore. We can't afford it anyway.	Reason 5:

4 Which one person do you think is the most likely to actually move? Say why.

_____

_____

5.6 Does geography help or hinder the Russian economy?

Assessing the worth of Yakutsk's transport network

Sakha, a Russian republic bigger than Argentina, produces 90 per cent of Russia's diamonds, 30 tons of gold a year and has oil and gas reserves. Its population is 900,000. Over 250,000 people live in Yakutsk, 4,884 km from Moscow. Russia's roads and railways are concentrated in the south and European west of the country.

Permafrost destroys the northern road surfaces in winter and covers them in mud/puddles in summer.

1 How far is it from London to your local town? _____ km.

2 How long does the journey take? By air _____   By rail _____   By car _____

3 Study the timetables below for journeys from Moscow to Yakutsk. How many hours does each journey take? Remember you need to take six hours off each arrival time to allow for the time difference.

Air Depart Monday 1800	Arrive Tuesday 0630	Time:		hours
Road Depart Monday 0900	Arrive Saturday 0700 (nonstop)	Time:	days	hours
Rail Depart Monday 0025	Arrive Sunday 2125	Time:	days	hours

4 Look at the list of 'good news' items about the transport situation in Yakutsk.

Draw a line to link 'good news' to its matching 'bad news' and 'worse news' about each item. The first is done for you.

GOOD NEWS	BAD NEWS	WORSE NEWS
Goods can be shipped in by river boat.	The repairs cost millions of roubles.	Without constant repair it will get potholes and mud-clogged again very quickly.
A rail link was completed in 2014.	The River Lena freezes in winter.	Winter is very long.
Cars can cross the river ice in winter. There is a ferry in summer.	In 2020, perhaps!	You can cross with a hydrofoil if you dodge the ice floes.
A bridge is being built.	The river is dangerous to cross in spring and autumn.	The bridge must be quite a distance upriver where the river narrows.
Food can be trucked in.	It's a two-week round trip by lorry.	There is no bridge across the river.
The Lena 'highway from hell' is now repaired and is passable all year.	The station is south of the river.	If you switch the engine off AT ANY TIME in winter you may freeze to death.

5 Do you think that maintaining the road and rail links to Yakutsk is worth the millions of roubles spent? YES / NO

6 Give at least one reason for your choice of answer. _____

_____

_____

5.7 What is GIS and how can I use it to investigate Russia?

The advantages of GIS images over normal mapping systems

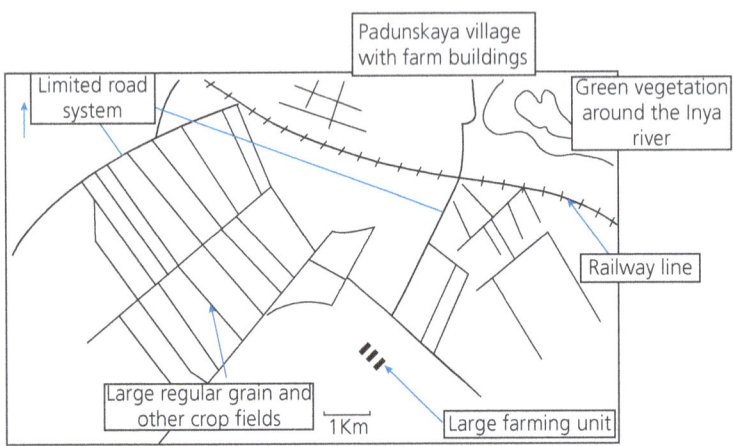

Map A: Sketch map of Padunskaya village, located between Novosibirsk and Kemerova

Study the sketch map of the Padunskaya area in southern Siberia, made from a Google image.

1 Which climatic area of Russia do you think it is in? Tick the correct answer.

The Tundra ☐   The Taiga ☐   The Steppe ☐

2 Give at least one reason why you think it is in this area. _____

_____

3 Investigate Yakutsk and its surroundings using Google maps and satellite imagery. Use the information to draw a sketch map in the box below. Label the R. Lena, Yakutsk city, ferry route, the Lena Highway, areas of forest and icy wetlands.

4 Explain how the satellite layer helped you add some of the features and labels to the map.

_____

_____

5 What extra data layers would help if you were …

  a) a policeman solving a crime in Yakutsk _____

  b) investigating pollution in the area _____

5.8 Why did Russia plant their flag on the seabed of the North Pole? Part 1

5.8

Factors influencing Russia's claim for the North Pole

1   Use the table below to draw a line graph to show how world gas use (consumption) has changed since 1980. Give your graph a suitable title.

Daily natural gas use 1980–2010 (cubic feet)

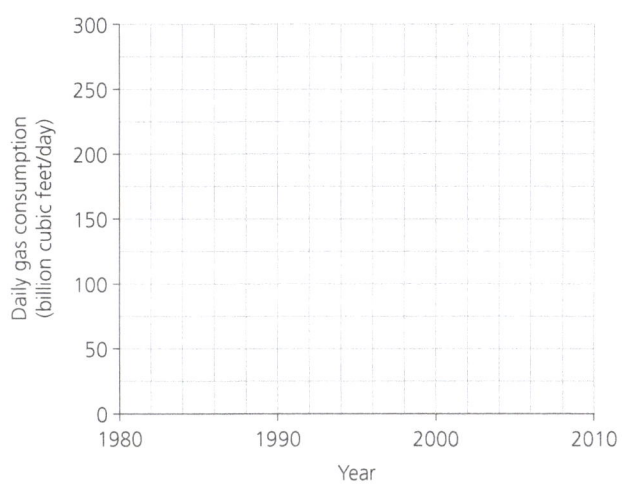

1980	1990	2000	2010
135 billion/ day	190 billion/ day	230 billion/ day	300 billion/ day

Natural gas is still measured in many different ways. This table uses billion cubic feet per day.

2   Complete the following sentences by circling the correct words: World gas consumption **rose / fell** between 1980 and 2010. It **more/less** than doubled in the **30 / 40** year period. There is a **great / low** demand to buy gas by many countries. Russia might want to claim an area of seabed rich in natural gas because it is **an economic asset / a military asset**.

3   Study the two pie charts below. What do they show about the percentages of gas used by Europe and Asia since 1980?

   a) The percentage of gas used by Europe has gone **up / down.**

   b) The percentage used by Asia has gone **up / down.**

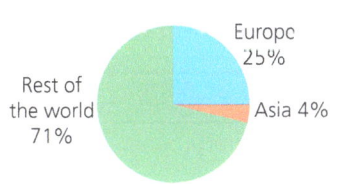

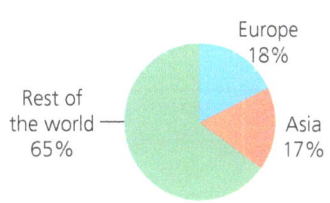

Pie chart A: World gas consumption 1980   **Pie chart B:** World gas consumption 2010

4   Which area is most likely to provide an expanding market for Russian gas?

_____

5   Give at least one reason that each of the following could help Russia to exploit and then sell the Arctic gas.

   a) Global warming means that _____

   _____

   b) More technologically advanced ships mean that _____

   _____

5.9

5.9 Why did Russia plant their flag on the seabed of the North Pole? Part 2

Gas exploration in the Tundra affects traditional lifestyles

The Nenets indigenous people's 600,000 reindeer must migrate 1,700 km north across the Yamal Peninsula to their summer feeding grounds by the Kara sea coast. South of the peninsula it is too warm and there are gnats. Reindeer herds can be enormous or quite small in number.

1 Find a picture of the Nenets' reindeer grazing in **summer** and study it.

2 Finish the sentence to explain how this challenges the view that all of the Arctic is an icy 'wasteland'.

This challenges the view that the Arctic region is always icy because _____

_____

3 Study the number of hazards the 600,000 reindeer have to encounter now the Yamal megaproject has developed the Bovanenkovo gas field across the migration route. They attempt to solve these problems and obstacles in a variety of ways. Complete the table using the hazards from the box. The last is completed for you.

A new railway   Busy trucking roads   Polluted lichen and grass   Pipelines at the surface

Gas field infrastructure/metal equipment   Building debris: scrap metal, glass

Hazard/obstacle	Solution
	Pipelines are raised at some critical points where they cross the migration route.
	Reindeer are supervised or enclosed for the night.
	Wooden boards are placed over the rails to protect the all-terrain sledge runners.
	Reindeer refuse to eat.
	Soft material is put down and the traffic stopped for up to 40 minutes.
Building debris: scrap metal, glass	The Nenets have to avoid or remove this.

4 Do you think the Nenets' lifestyle and their reindeer migration will survive in the future? Circle your opinion: **YES / NO / POSSIBLY / DOUBTFUL**

Justify your answer.

_____

_____

© Hodder Education 2018

5.10 Is the geography of Russia a curse or a benefit? Review

5.10

Student's Book
pages 122–123

1 Read the following statements. Use a tick to indicate whether you agree, disagree or partly agree with each of them. Number 4 has been completed for you.

	Agree	Disagree	Partly agree
1 Russia is a vast country with lots of mineral wealth.			
2 Russia's landscape is very varied and never boring.			
3 Russia has low levels of energy reserves like oil, gas and coal.			
4 All of the northern region should be left untouched.			✓
5 Russia's minerals should make it more wealth than they do.			
6 Climate does not really limit Russia's agriculture or mining profits.			
7 Russia's population is too large for the area of the country.			
8 The population density is even throughout the country.			
9 Russian people are very hardy.			
10 Distance is an enormous barrier for Russian development.			
11 The road and rail system in not sufficient to help Russia's economy grow.			
12 The money spent on the transport system is completely wasted.			
13 All Russians have equal access to schools and medical care.			
14 Global warming is helping Russia's economy.			
15 Russia's oil and gas exploitation endangers the environment and the nomadic herders.			
16 Russia's claim to part of the north polar area is not fair.			

2 Chose a statement you only partly agree with and explain why you think it is partly correct and why you think it is partly wrong.

_____

_____

_____

_____

_____

_____

_____

_____

_____

_____

PROGRESS IN

WORKBOOK 1

GEOGRAPHY

UNITS 1–5

KEY STAGE 3

Review and reinforce the skills, knowledge and understanding that you are developing throughout your Progress in Geography: Key Stage 3 course.

This Workbook accompanies your Student Book, providing extra support as you continue on your journey to become a good geographer.

- A range of activities focus on skills, knowledge and understanding
- Ideal for homework, classwork and independent study
- One Workbook page for every lesson in the Student Book

Also available:

Workbook 1: Units 1–5 (Pack of 10)
ISBN: 9781510442993

Workbook 2: Units 6–10 (Single copy)
ISBN: 9781510428065

Workbook 2: Units 6–10 (Pack of 10)
ISBN: 9781510443006

Workbook 3: Units 11–15 (Single copy)
ISBN: 9781510442986

Workbook 3: Units 11–15 (Pack of 10)
ISBN: 9781510443013

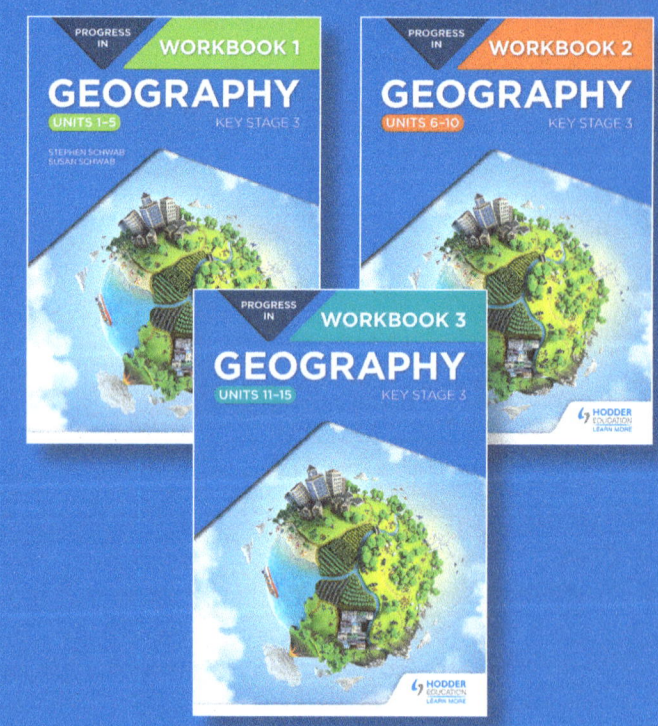

HODDER EDUCATION

t: 01235 827827
e: education@hachette.co.uk
w: hoddereducation.co.uk

ISBN 978-1-5104-2807-2

MIX
Paper from
responsible sources
FSC™ C104740

PROGRESS IN

WORKBOOK 1

GEOGRAPHY

Units 1–5

KEY STAGE 3

STEPHEN SCHWAB
SUSAN SCHWAB

HODDER
EDUCATION
LEARN MORE

NAME: ....

CLASS: ....

The Publishers would like to thank the following for permission to reproduce copyright material.

Photo credits

p.2 l © Old Paper Studios/Alamy Stock Photo; r NASA Goddard Space Flight Center Image by Reto Stöckli (land surface, shallow water, clouds). Enhancements by Robert Simmon (ocean color, compositing, 3D globes, animation). Data and technical support: MODIS Land Group; MODIS Science Data Support Team; MODIS Atmosphere Group; MODIS Ocean Group Additional data: USGS EROS Data Center (topography); USGS Terrestrial Remote Sensing Flagstaff Field Center (Antarctica); Defense Meteorological Satellite Program (city lights); **p.10, 13** © Stephen Schwab, Susan Schwab.

Acknowledgements

Every effort has been made to trace all copyright holders, but if any have been inadvertently overlooked, the Publishers will be pleased to make the necessary arrangements at the first opportunity.

Although every effort has been made to ensure that website addresses are correct at time of going to press, Hodder Education cannot be held responsible for the content of any website mentioned in this book. It is sometimes possible to find a relocated web page by typing in the address of the home page for a website in the URL window of your browser.

Hachette UK's policy is to use papers that are natural, renewable and recyclable products and made from wood grown in well-managed forests and other controlled sources. The logging and manufacturing processes are expected to conform to the environmental regulations of the country of origin.

Orders: please contact Hachette UK Distribution, Hely Hutchinson Centre, Milton Road, Didcot, Oxfordshire, OX11 7HH. Telephone: +44 (0)1235 827827. Email education@hachette.co.uk Lines are open from 9 a.m. to 5 p.m., Monday to Friday. You can also order through our website: www.hoddereducation.co.uk

ISBN: 9781510428072

First published in 2018 by

Hodder Education,

An Hachette UK Company

Carmelite House

50 Victoria Embankment

London EC4Y 0DZ

www.hoddereducation.co.uk

Impression number   10 9 8

Year   2023

Cover photo ©arquiplay77 - stock.adobe.com

Illustrations by Integra

Typeset in India by Integra

Printed in Spain by Graphycems

A catalogue record for this title is available from the British Library.